Structured Epistemic Truth – Volume I

Foundations, Perspectives, and Shifting Paradigms

Krister Sundholm

May 26, 2025

Contents

Chapter 1

Preface

This volume is the result of several years of reflection on the nature of mathematical truth, epistemic access, and the limits of formalism. The Structured Epistemic Model of Truth (SEMT) emerged not as a finished theory, but as a scaffolding — an architectural lens for navigating questions that have long resisted resolution.

This book is written for mathematicians, philosophers, logicians, and anyone interested in the foundations of knowledge. It aims to be rigorous without being rigid, and ambitious without overreaching. It offers not finality, but structure — and structure, here, is the beginning of clarity.

I am grateful to the many conversations, silences, paradoxes, and failures that shaped this work. Truth, as SEMT teaches us, does not always arrive as a proof. Sometimes it appears as a structure unfolding over epistemic time.

Chapter 2

How to Read This Volume

This volume is organized into three parts:

- **Part I: The Epistemic Turn** introduces the four SEMT layers of truth — objective, inductive, deductive, and indeductive. Each chapter defines, motivates, and distinguishes these levels.
- **Part II: Structural Explorations** applies the SEMT framework to longstanding problems in logic and computation, such as the Continuum Hypothesis and P vs NP.
- **Part III: Implications and Outlook** reflects on SEMT's philosophical consequences and outlines a roadmap for future research.

Readers may choose to follow the book linearly or jump to sections of interest. Philosophers may begin with the final chapters; mathematicians may dive directly into Part II. However, the full arc emerges only when the structure is taken as a whole.

SEMT is not only a theory to be studied, but a lens through which we may re-see familiar territory. Read with structure in mind.

2.1 What is SEMT?

The Structured Epistemic Model of Truth (SEMT) is a layered framework for understanding mathematical truth in terms of epistemic access and structural validation. Unlike traditional truth theories, which aim to describe truth as a singular or fixed notion, SEMT embraces a pluralistic, evolving architecture. It defines truth not just as provability, but as a position within a hierarchy of epistemic processes.

This chapter introduces SEMT's purpose: to clarify where and why classical foundations reach their limits, and to propose an alternative based not on logical reduction, but on structured epistemic ascent.

Part I

The Epistemic Turn

Chapter 3

Objective Truth

3.1 Introduction

Objective truth, in the context of Structured Epistemic Truth (SEMT), refers to truths that are epistemically grounded in finite verification. This is the foundational layer of the SEMT hierarchy, representing the most constrained but also the most accessible level of truth within a computational or logical system. It captures truths that can be directly observed, calculated, or confirmed through bounded, resource-finite processes.

3.2 Motivation

The classical notion of truth as correspondence — most prominently captured in Tarski's semantic theory — presupposes a metaphysical structure in which truth is a relation between statements and objective reality. SEMT retains the clarity of this concept but relocates it within the bounds of epistemic accessibility. That is, a truth is objective when it can be established within a finite, rule-bound structure available to a bounded agent.

In computational terms, this corresponds to statements that can be verified within polynomial time or via deterministic algorithms of fixed complexity. In formal logic, it parallels statements provable within a weak but complete fragment of arithmetic or set theory.

3.3 Definition of Objective Truth

Objective truth in SEMT is defined as truth validated by direct, finite procedures under fixed epistemic conditions. This includes:

- Finitely checkable proofs
- Decidable propositions in arithmetic
- Truths verifiable by a single Turing machine within bounded resources

Importantly, objective truth does not imply absoluteness — only that, relative to a bounded agent or machine, the proposition can be confirmed without induction, abstraction, or indeductive unfolding.

3.4 Comparison with Classical Frameworks

Tarski's framework depends on meta-languages and infinite sets. Objective truth avoids this by remaining fully grounded in internal epistemic operations. It aligns more with early formalist and intuitionist positions but offers a structured hierarchy rather than a binary dichotomy between provable and unprovable.

In contrast to Kripke's fixed-point approach to truth (which tolerates semantic paradoxes), SEMT's objective layer tolerates only those statements that can be grounded without recursion or circularity.

3.5 Epistemic Character

The defining epistemic characteristic of objective truth is stability under immediate verification. That is:

> If a statement is objectively true, then any bounded agent operating under the same rules will converge on its truth value without additional inference.

3.6 Limitations and Boundaries

Objective truth cannot capture statements whose verification depends on:

- Infinite processes

- Probabilistic inference
- Meta-theoretic consistency arguments
- Long-term epistemic convergence

Thus, while foundational, it is insufficient for expressing deeper logical truths or for resolving statements such as the Continuum Hypothesis or the P vs NP problem.

3.7 Role within SEMT

Objective truth serves as the "base layer" in SEMT — the level upon which all others depend but from which they epistemically diverge. It provides grounding, stability, and immediate checkability.

Its primary contribution is to demonstrate that some truths require no abstraction or recursion — they are already within reach, as long as the structure verifying them is adequately formalized and resource-bounded.

3.8 Transition to Higher Levels

From here, SEMT expands upward into inductive truth — where stabilization across models replaces immediate verifiability. This transition marks the first step away from finite methods and toward epistemic generalization.

In SEMT, each level is not a rejection of the previous, but a generalization beyond its constraints. Thus, the next chapter introduces inductive truth — truth as asymptotic agreement across admissible structures.

Chapter 4

Inductive Truth

4.1 Introduction

Inductive truth in SEMT represents the second epistemic layer — one that transcends direct verification and enters the realm of stabilization across epistemic space. Where objective truth is rooted in the certainty of bounded agents and immediate operations, inductive truth is about convergence: not what is instantly known, but what eventually becomes known through repeated, structure-relative validations.

4.2 Motivation

Many truths in mathematics and logic resist finite verification but exhibit a stable pattern across various admissible models. For example, certain number-theoretic conjectures may not be provable in ZFC, yet hold across all known consistent models. Inductive truth captures this idea epistemically: truth as persistent reappearance, not absolute derivability.

This reflects the philosophical insight that knowledge may emerge not from a single proof, but from the persistent absence of counterexamples across a landscape of systems. It also corresponds to informal mathematical practice — where we accept statements as true after sustained consistency across independent formulations.

4.3 Definition of Inductive Truth

Inductive truth is defined in SEMT as truth stabilized across a family of epistemically admissible structures. These structures may vary in formulation, strength, or axiomatization, but if a proposition holds across them without contradiction, it earns inductive status.

Formally, let $\mathcal{F}$ be a family of admissible models M_i. A statement ϕ is inductively true if:

> ϕ holds in all $M_i \in \mathcal{F}$, for $\mathcal{F}$ closed under epistemic expansion.

This formulation introduces convergence as a new epistemic criterion — different from proof or computation.

4.4 Comparison with Classical Frameworks

Inductive truth is not reducible to provability. In classical logic, truth is model-relative; in SEMT, it is model-stabilized. This echoes the shift from Gödel's syntactic truth to semantic resilience.

It also extends ideas from Kripke's theory of truth, though SEMT's inductive layer avoids semantic paradoxes by requiring cross-model convergence, not self-referential closure.

4.5 Epistemic Character

The core epistemic feature of inductive truth is resistance to falsification. While objective truth relies on confirmation, inductive truth arises from an absence of contradiction — across epistemic space rather than within a single structure.

> If a statement is inductively true, no bounded countermodel disproves it within any admissible extension.

4.6 Limitations and Boundaries

Inductive truth cannot distinguish between statements that are:

- Universally true vs. empirically persistent
- Logically unprovable but structurally stable
- Reflectively true vs. epistemically undecidable

It also depends on how one defines "admissible" models — a subjectively determined boundary within SEMT.

4.7 Role within SEMT

Inductive truth generalizes objective truth. It retains a grounding in structure, but no longer insists on finiteness or immediacy. It introduces the idea that truth can be *distributed epistemically*, not just located.

This layer opens the door to handling the Continuum Hypothesis and other undecidable statements — not by resolving them, but by reframing them within an epistemic model of convergence.

4.8 Transition to Higher Levels

The next level — deductive truth — moves from convergence to coherence. If inductive truth is structural agreement, deductive truth is inferential integrity. It is no longer enough to see stability; now we seek a system where that stability arises from internal rules of derivation.

Chapter 3 will explore how SEMT frames deductive truth as the closure of epistemic structure under self-consistent inference.

Chapter 5

Deductive Truth

5.1 Introduction

Deductive truth forms the third epistemic layer in SEMT. It shifts the focus from observed stability across models (inductive truth) to internally generated stability within a system of inference. Where inductive truth identifies statements that hold across a range of admissible structures, deductive truth emerges from inferential closure: truth preserved and derived from within a coherent, rule-governed system.

5.2 Motivation

Mathematics, as practiced, often hinges not just on the recurrence of stable patterns, but on systems of deduction that support expansive inferential chains. Deductive truth captures the kind of epistemic legitimacy that arises when a statement is not just widely accepted, but inferable from shared assumptions using valid inference rules.

This level reflects the transition from external epistemic validation to internal epistemic construction — where truth is not only stable, but structurally necessary given the system's logic.

5.3 Definition of Deductive Truth

A proposition is deductively true within SEMT if it follows from a set of axioms and inference rules under a system that is itself internally coherent and epistemically grounded.

Let Σ be a set of premises, and R a set of admissible inference rules. Then a proposition ϕ is deductively true if:

> $\Sigma \vdash_R \phi$ and (Σ, R) form a consistent, epistemically admissible deductive structure.

Unlike classical deduction, which is defined relative to abstract proof systems, SEMT imposes epistemic constraints on both the rules and the premises — ensuring that the system itself is not merely formal, but also structurally reliable.

5.4 Comparison with Classical Frameworks

SEMT's notion of deductive truth refines classical proof-theoretic approaches by embedding them in an epistemic hierarchy. In contrast to Hilbert-style systems or natural deduction, SEMT does not treat all derivations equally — only those within epistemically stable systems count.

This idea shares affinity with constructivist logics, but SEMT is not bound to intuitionistic rejection of excluded middle. Instead, it filters deduction through an epistemic lens, not a metaphysical one.

5.5 Epistemic Character

Deductive truth in SEMT is defined by inferential necessity. Once a proposition enters the space of deductive truth, it is no longer validated by stability or observation — but by its position in a web of internal implication.

> If a statement is deductively true, then within the coherent structure of its system, it cannot be otherwise.

This sense of "cannot be otherwise" marks the transition from verification (objective) and convergence (inductive) to implication and necessity.

5.6 Limitations and Boundaries

Deductive truth depends on:

- The admissibility of the underlying axioms
- The epistemic soundness of inference rules
- The internal consistency of the system

Thus, it cannot account for truths that are externally consistent but internally unprovable. Nor can it resolve questions whose answers require appeal beyond the formal system.

5.7 Role within SEMT

Deductive truth sits above inductive truth, not in authority, but in generality. It permits internal derivation of results that inductive truth merely accepts via cross-model agreement.

It models the logical stage of epistemic activity — where agents reason, derive, and extend knowledge beyond what they can observe or verify externally.

5.8 Transition to Higher Levels

The next chapter introduces indeductive truth — the highest layer in SEMT — which reframes truth not as a static property of systems or models, but as a dynamic feature of epistemic time. It allows for truths that emerge only through the unfolding of structure over time, even if not derivable or immediately convergent.

Chapter 6

Indeductive Truth and Epistemic Time

6.1 Introduction

Indeductive truth represents the fourth level in SEMT's epistemic hierarchy. It addresses truths that are not grounded in immediate verification, cross-model stability, or internal deduction — but instead unfold over epistemic time. These are truths that emerge only after epistemic processes mature, and whose recognition depends on temporal patterns, not just structural properties.

6.2 Motivation

In both mathematics and cognition, there are propositions that seem inaccessible at any fixed stage, but whose status becomes apparent across sequences of reasoning, reflection, or theory evolution. This is especially relevant for unresolved problems like the Continuum Hypothesis or P vs NP, where no stable proof or falsification has emerged, yet the epistemic discourse evolves.

SEMT models these cases not as failures of truth, but as truths that are epistemically deferred — whose emergence is conditional on long-term structural unfolding.

6.3 Definition of Indeductive Truth

A proposition is indeductively true if it stabilizes in the limit across epistemic stages indexed by a structured notion of time. That is, its truth is defined not at a single moment, but as an asymptotic property of an evolving epistemic process.

Let $T_0, T_1, T_2, \ldots$ be a sequence of epistemic stages. Then a proposition ϕ is indeductively true if:

> There exists $n \in \mathbb{N}$ such that for all $m \geq n$, ϕ is accepted as true in T_m, and this stability is not reversed.

This defines truth as convergence in epistemic time — analogous to the way limits define values in analysis.

6.4 Comparison with Classical Frameworks

Indeductive truth has no analogue in classical logic, which is timeless and static. It overlaps conceptually with ideas in belief revision, learning theory, and modal epistemology — but SEMT formalizes it as a truth level, not just a model of inquiry.

It also resonates with intuitionistic delay, but without rejecting classical logic. SEMT reframes delay as a structural, not subjective, epistemic effect.

6.5 Epistemic Character

Indeductive truth is defined by irreversible stabilization over time. Its epistemic profile includes:

- Delayed knowability
- Irreversible convergence
- Model-relative unfolding

> If a statement is indeductively true, then though not accessible now, it will stabilize epistemically in the long run.

This adds a dynamic dimension to truth — where status evolves rather than being fixed.

6.6 Limitations and Boundaries

Indeductive truth does not guarantee provability or present knowability. It may remain inaccessible at any finite point.

It also depends on the structure of epistemic time — which is itself a conceptual choice. Thus, different definitions of epistemic time may yield different stabilizations.

6.7 Role within SEMT

Indeductive truth completes the SEMT hierarchy by capturing truths that are invisible to earlier layers. It handles cases where classical tools fail, yet epistemic evolution reveals consistent, irreversible tendencies.

In this way, SEMT offers a framework for engaging with undecidable propositions not by forcing resolution, but by modeling their epistemic trajectory.

6.8 Concluding Reflection

With the Indeductive layer, SEMT establishes a full epistemic spectrum: from immediate verification to infinite stabilization. Each level generalizes the previous, offering a lens to reinterpret foundational questions not through metaphysics, but through epistemic architecture.

The next part of the book will explore how this architecture restructures longstanding questions in logic and computation.

Part II

Structural Explorations

Chapter 7

Truth Without Proof

7.1 Introduction

This chapter initiates the second part of SEMT by shifting focus from the definition of epistemic levels to how they reshape foundational problems in logic and mathematics. We begin with a provocative theme: truth without proof. SEMT enables us to understand how certain truths may emerge and stabilize epistemically, even when no formal proof exists within accepted deductive systems.

7.2 Motivation

There are well-known mathematical propositions, such as the Continuum Hypothesis (CH) or certain large cardinal statements, for which formal proof or disproof within ZFC is currently impossible. Rather than interpret this as a failure, SEMT treats these cases as signals of epistemic evolution. It offers an alternative framework in which truth is not equivalent to proof, but may emerge through layered epistemic convergence.

This opens the door to rethinking proof not as a binary gate to truth, but as one epistemic path among others.

7.3 Structural Epistemic Perspective

In SEMT, a statement can be objectively unprovable but inductively or indeductively true. "Truth without proof" is thus reframed not as an anomaly, but as a meaningful epistemic category. When a statement persists across extended models or stabilizes over epistemic time, its status becomes more than undecidable — it becomes structurally significant.

7.4 Case Types

- Statements holding across all known models, but unprovable: candidates for inductive truth
- Statements with high-level informal acceptance but no formalization: candidates for deductive emergence
- Statements resistant to formal treatment but stable over time: candidates for indeductive truth

SEMT turns these categories into a working epistemic grammar.

7.5 Epistemic Implications

The recognition of truth without proof pushes mathematics away from static formalism and toward epistemic pluralism. SEMT does not deny the value of proof — it embeds proof within a broader hierarchy of justification. This allows working mathematicians and philosophers to interpret their intuitions, conjectures, and informal insights within a principled framework.

7.6 Role in SEMT

This chapter marks a key shift: we move from defining truth layers to exploring their consequences. Truth without proof is not a contradiction — it is an invitation to epistemic maturity. In SEMT, such truths are located, analyzed, and sometimes elevated — even if they remain beyond derivation.

7.7 Transition to Continuum Hypothesis

The next chapter takes up one of the most famous cases of truth without proof: the Continuum Hypothesis. We will examine how SEMT reframes CH, not by solving it, but by revealing its epistemic character — and placing it within the layered dynamics of structured truth.

Chapter 8

Reframing the Continuum Hypothesis

8.1 Introduction

The Continuum Hypothesis (CH) is a landmark problem in the history of logic. It asks whether there exists a set whose cardinality lies strictly between that of the integers and the real numbers. Since the work of Gödel and Cohen, we know that CH is independent of the standard axioms of set theory (ZFC). SEMT offers a new perspective: CH is not a problem of missing proof, but a case study in epistemic layering.

8.2 Motivation

CH has resisted resolution for nearly a century not because it is inherently unknowable, but because its truth status is not absolute in the classical sense. SEMT interprets this resistance not as a failure, but as a signal — that CH is best understood at a higher epistemic level, where stability, not derivability, determines its truth.

8.3 SEMT Reframing

Rather than seeking a proof or disproof of CH, SEMT asks: how does CH behave across epistemic space? Is it stable in models that are admissible under certain epistemic conditions? Does it recur at the inductive or indeductive level?

Under this approach, CH becomes a structurally analyzable phenomenon:

- At the objective level: undecidable
- At the inductive level: exhibits model-sensitive divergence
- At the Indeductive level: candidate for eventual epistemic stabilization

8.4 Philosophical Implications

By reframing CH, SEMT avoids metaphysical deadlock. Instead of insisting that CH has a determinate truth value in some Platonic universe, SEMT sees it as an epistemic variable whose truth character emerges only within a structured hierarchy. This aligns with the philosophical view that mathematical truth is not static, but layered and evolving.

8.5 A Structural Epistemic Diagnosis

SEMT offers a new diagnosis: CH appears unsolvable because it is being forced into a truth layer that cannot accommodate it. By placing CH in the context of indeductive truth, we acknowledge that some mathematical questions only become meaningful over time — not through deduction, but through structural convergence.

8.6 Broader Lessons

CH is not an exception. It is a prototype. SEMT uses it to illustrate how epistemic layering can reclassify longstanding problems without reducing them. Other statements (e.g. large cardinals, projective determinacy) may follow similar trajectories.

8.7 Transition to Complexity

The next chapter will move from set-theoretic independence to computational asymmetry. The question of P vs NP is not formally analogous to CH, but epistemically similar: it may resist resolution at one layer while stabilizing at another. SEMT offers a lens through which such asymmetries become not failures, but structural features.

Chapter 9

Complexity and Incompleteness

We introduce the term *indeductive* to describe Level 2.5 of SEMT — a structured epistemic layer between inductive convergence and deductive formalism. This level captures truths that stabilize over time through partial formalization, emergent regularity, or temporal convergence that is not yet fully axiomatic.

9.1 Introduction

This chapter bridges the epistemic layering of SEMT with complexity theory and incompleteness. Further examines how certain problems, notably P vs. NP, resist resolution not merely for technical reasons, but because of deep structural asymmetries in the epistemic framework. SEMT reinterprets these issues as signals of misalignment between epistemic levels.

9.2 Motivation

Traditional complexity theory frames the P vs. NP problem in terms of computational classes: whether problems whose solutions can be verified quickly can also be solved quickly. But this framework ignores the epistemic structure underlying the distinction between 'solution' and 'verification'. SEMT exposes that the divide P vs NP may reflect a deeper truth asymmetry - one not captured by Turing-style models alone.

9.3 Asymmetric Epistemic Layers

SEMT introduces a key insight: solution and verification are not symmetric activities across all truth layers. In particular:

- At the objective level, both are reducible to bounded computation.
- At the inductive level, verification stabilizes more easily than solution.
- At the indeductive level, solutions may emerge only after long epistemic unfolding, while verification can remain bounded.

This structural asymmetry may be the real source of the P vs NP gap.

9.4 SEMT Perspective on Incompleteness

Incompleteness theorems are typically interpreted syntactically: statements true but unprovable within a system. SEMT reframes them as epistemic boundaries between layers:

- Gödel sentences become indeductively true but deductively unreachable.
- Reflection principles live at inductive levels but not within base systems.

SEMT therefore naturalizes incompleteness as a feature of epistemic architecture, not a limitation of formalism.

9.5 P vs NP as Epistemic Tension

P vs NP now appears not as a technical problem alone, but a misclassification of epistemic layers:

- NP verification fits within bounded objective truth.
- P-style solving often involves broader structural insight — potentially inductive or indeductive.

Thus, SEMT implies that the question may be ill-posed if asked within a single-layer epistemology.

9.6 Transition to Structural Truth

The next chapter addresses SEMT as a structural theory of truth. What emerges from these layered insights is a coherent epistemic framework — one that not only accommodates complexity and independence, but reclassifies them as natural outcomes of layered truth.

Chapter 10

SEMT as a Structural Theory of Truth

10.1 Introduction

This chapter consolidates the conceptual structure developed throughout the preceding chapters. SEMT, having shown its capacity to reframe undecidability, computational asymmetry, and formal independence, now emerges as a candidate for a general theory of structured truth — not just a framework, but a paradigm.

10.2 From Layers to System

Each SEMT layer — objective, inductive, deductive, and indeductive — has been treated individually. Now we examine their interaction. SEMT is not simply a taxonomy; it is a structural system in which truths flow, evolve, and stabilize across epistemic thresholds.

- Objective truth provides immediate grounding.

- Inductive truth extends through cross-model convergence.
- Deductive truth organizes and integrates within formal systems.
- Indeductive truth projects forward across epistemic time.

Together, they define a layered, recursive structure — not linear, but dynamically scaffolded.

10.3 Reframing Truth Itself

Traditional accounts of truth — correspondence, coherence, pragmatist — emphasize different aspects. SEMT incorporates these while adding epistemic structure as a central organizing principle. In SEMT:

> Truth is a property of statements relative to an epistemic system, evolving through layers of admissibility, inference, and stability.

This turns truth into a structured process, not a binary predicate.

10.4 Implications for Logic and Foundations

By elevating epistemic architecture, SEMT invites a shift in logic itself. We no longer judge a logical system by completeness alone, but by how it interacts with epistemic boundaries. This offers a new interpretation of independence, inconsistency, and expressivity.

10.5 Implications for Computation

SEMT suggests that complexity theory, especially P vs NP, reflects an epistemic layering issue rather than a computational deadlock. It reframes "hard problems" as cases of epistemic misclassification — urging the development of complexity models sensitive to structural asymmetry.

10.6 The SEMT Paradigm

SEMT proposes a new epistemic lens for mathematics, logic, and computation. It is not a replacement for existing methods, but a meta-framework that:

- Locates truth relative to structured knowledge processes
- Reclassifies classical dilemmas as layer-crossing challenges
- Invites new forms of analysis, comparison, and epistemic clarity

This marks the transition from truth as a static property to truth as a layered epistemic phenomenon.

10.7 Transition to Part III

Having established SEMT as a structural theory of truth, we now turn to broader implications: how SEMT interacts with real mathematical practice, philosophical foundations, and the future of epistemic modeling. Part III expands the framework into its philosophical and methodological reach.

Part III

Relevance, Reach, and Reserve

Chapter 11

$P \neq NP$ as Epistemic Asymmetry

11.1 Introduction

This chapter reexamines the P vs NP problem not from a purely computational or formal perspective, but from an epistemic one. SEMT enables us to recast $P \neq NP$ as a structural asymmetry between layers of truth, rather than a technical or unresolved question within a single model. This shift exposes new dimensions of the problem — and suggests why classical tools may be epistemically misaligned with the nature of the divide.

11.2 Motivation

The question "Is P equal to NP?" appears deceptively simple: can every problem whose solution can be verified in polynomial time also be solved in polynomial time? Yet decades of effort have failed to resolve it. SEMT offers an explanation: this problem lives across epistemic layers that operate under different conditions for truth — and this mismatch is what gives rise to the problem's apparent intractability.

11.3 Verification vs. Solution as Layered Acts

Within SEMT, verification and solution represent different epistemic behaviors:

- Verification typically belongs to objective or low inductive layers — bounded, rule-checkable.
- Solution often depends on inductive discovery or even indeductive stabilization — a higher epistemic demand.

This reframing reveals that the P vs NP distinction is not merely about resource bounds but about the nature of epistemic access.

11.4 Epistemic Asymmetry in Practice

SEMT highlights a core asymmetry:

> In many systems, verifying a truth is epistemically easier — and categorically lower — than constructing it.

This gap is not incidental but structural. It mirrors the layer gaps in SEMT:

- From indeductive truth to objective verification: an epistemic descent
- From stable convergence to checkable outcome: a compression of epistemic scope

The asymmetry is not an error in the model — it is the model.

11.5 Consequences for the Classical Problem

If $P \neq NP$ is a mismatch of layers, then attempting to resolve it within a single-layer framework (e.g. classical Turing computation) may be epistemically blocked. SEMT does not prove $P \neq NP$ in the traditional sense — it reframes the question:

> Is the epistemic act of solving the same kind of act as verifying?

Under SEMT, the answer is: no. Solving may require epistemic ascent that verification does not.

11.6 Implications for Proof and Strategy

This reframing explains why proof efforts have stalled: they often rely on collapsing both acts into the same truth layer. SEMT warns against this. Instead, it invites new strategies that respect the layer distinction:

- Model solution processes across epistemic time
- Treat verification as boundary condition, not total structure
- Recast search and discovery as inductive-indeductive hybrids

11.7 Beyond P vs NP

The P vs NP divide may be the most visible case, but it's not the only one. Many "hard" problems exhibit the same epistemic asymmetry. SEMT could provide a systematic framework for identifying and classifying them — not by computational hardness, but by their location in epistemic space.

11.8 Transition to Philosophical Implications

This chapter concludes the structural application of SEMT. What follows is a broader philosophical synthesis: what it means to do epistemic mathematics, to theorize truth without absolutes, and to model knowledge not as static facts but structured layers in motion.

Chapter 12

Philosophical Implications of SEMT

12.1 Introduction

The Structured Epistemic Model of Truth (SEMT) is not merely a technical innovation or a response to classical problems. It is a philosophical act: a reframing of what it means to know, to prove, and to declare something as true. In this chapter, we explore how SEMT disrupts traditional metaphysical and epistemological views, and what it offers as a new foundation for thinking about mathematics, logic, and cognition.

12.2 Truth Without Absolutes

SEMT challenges the long-standing assumption that truth must be singular, absolute, and context-free. In contrast, it positions truth as a dynamic, stratified phenomenon — layered according to epistemic access and validation mechanisms. This aligns with constructivist intuitions but formalizes them without discarding classical tools.

> Truth, in SEMT, is not weakened by structure.
> It is empowered by it.

Each SEMT layer retains its integrity, but gains significance through its position in the hierarchy. This provides a middle ground between relativism and foundationalism — a structured pluralism of truth.

12.3 Revisiting Logical Realism

Logical realism holds that logic reflects the structure of reality. SEMT reinterprets this by suggesting that logic reflects the structure of our epistemic operations. We do not discover logic as a metaphysical mirror; we construct and navigate it as a scaffold for layered understanding. This calls for a philosophical shift from ontology to epistemic architecture.

12.4 Epistemic Humility and Ascent

SEMT encourages a kind of epistemic humility. The model acknowledges that many truths may be inaccessible within our current systems — not because they do not exist, but because our tools have not reached the appropriate layer.

> Truth is not blocked by undecidability. It is deferred until the epistemic structure matures.

This humility is paired with an invitation to ascend. SEMT opens the door to higher-level reasoning where temporal convergence and structural stability reveal what proof cannot.

12.5 Implications for Mathematical Practice

SEMT does not reject formalism, platonism, or constructivism. It absorbs them. Each becomes a special case of a layered epistemology:

- Formalism aligns with objective and deductive layers
- Constructivism echoes inductive convergence
- Indeductive truth as a conceptual model for extending platonism

Mathematics, under SEMT, is a practice of epistemic exploration — not merely theorem-proving, but the strategic navigation of truth layers.

12.6 From Proof to Process

In SEMT, proof is one epistemic process among many. This reframes mathematical rigor not as fixation on derivability, but as discipline in understanding how truths migrate, stabilize, or resist collapse across layers. SEMT calls for new forms of justification, narrative, and methodological awareness in formal work.

12.7 Truth as Structured Insight

Ultimately, SEMT transforms the notion of truth into an insight conditioned by structure. A statement is true not in virtue of some metaphysical fact alone, but by virtue of its position within an evolving epistemic system.

> To know is not only to prove — it is to locate within a structure of unfolding insight.

12.8 Toward a SEMT Philosophy

The SEMT paradigm suggests the need for a broader epistemology — one that treats knowledge as layered, reflexive, and temporally rich. Such a philosophy can guide not only mathematics, but fields like AI, decision theory, and cognitive science, where epistemic structure is often implicit but under-theorized.

12.9 Transition to SEMT Roadmap

We now turn to the final chapter of this volume — a roadmap for SEMT. What questions does it unlock? What work remains? And how might it reshape both philosophical and technical landscapes in the years to come?

Chapter 13

The SEMT Roadmap

13.1 Introduction

This final chapter outlines the path forward. SEMT, as a model and a paradigm, opens a field — not just a perspective. The work presented here is foundational, but incomplete. In what follows, we chart its most immediate extensions, its long-term goals, and the strategic opportunities it offers for mathematics, philosophy, logic, and epistemology.

13.2 Open Questions

- How can SEMT be formalized beyond conceptual layers?
- What kinds of logic or semantics are needed to describe layer transitions?
- Can we construct explicit systems that demonstrate SEMT behavior?

- Is there a canonical mapping from classical truth theories to SEMT levels?
- What constraints define the admissibility of temporal stabilization?

These questions invite new logics, new proof systems, and new conceptual tools.

13.3 Theoretical Directions

- **SEMT Logic:** Development of multi-level formal systems with layered inference rules
- **SEMT Semantics:** Formal definitions of model stability, convergence, and admissibility
- **Epistemic Metrics:** Quantitative and structural criteria for truth migration across layers

These directions offer a scaffolding for extending SEMT into more formal, testable domains.

13.4 Applied Directions

- **P vs NP:** Explore verifier/solver asymmetries in explicit epistemic machine models
- **Undecidability:** Reclassify unresolved problems through SEMT's lens
- **Proof Theory:** Build formal calculi that reflect layer transitions

These projects serve both foundational insight and real technical benefit.

13.5 Strategic Trajectory

SEMT is more than a philosophical position. It is a proposed shift in how truth is modeled and handled. The strategic trajectory includes:

- Establishing SEMT as a legitimate subfield of logic and epistemology
- Publishing the foundational theory (this volume) and follow-ups (applied and formal volumes)
- Engaging with researchers in logic, CS, philosophy of mathematics, and foundations

This roadmap gives SEMT an identity — and a future.

13.6 A Paradigm in Formation

SEMT will evolve. This volume does not close the subject; it opens the possibility space. The goal is not to dominate, but to clarify — not to replace, but to restructure. In the end:

> SEMT is not a theory about truth. It is a theory of how we structure knowing.

13.7 Closing Note

To the reader, the theorist, the explorer of structure: this is your invitation. SEMT is now a landscape. What you do within it — that will determine its future.

Chapter A

Glossary of SEMT Terms

SEMT Structured Epistemic Model of Truth — a layered epistemic framework for truth in mathematics and logic.

Objective Truth Truth verifiable by bounded, finite methods.

Inductive Truth Truth stabilized across a family of admissible models.

Indeductive Truth Truth stabilized over epistemic time — emerging through long-term convergence.

Deductive Truth Truth derived from internal, coherent inference systems.

Epistemic Layer A level of structured validation within SEMT.

Truth Migration The movement of propositions between layers as structure unfolds.

Truth Layer	Verifiable	Provable	Stable	Emergent
L_1 Objective	Yes	Sometimes	Yes	No
L_2 Inductive	Model-relative	Not always	Yes	No
$L_{2.5}$ Indeductive	Not directly	Rarely	Eventually	Yes
L_3 Deductive	Internally derived	Yes	Yes	No

Chapter B

Author's Note

The ideas in this book evolved over time, and so did the language. Earlier terms, such as "Sundholm structure" or "semantic layering," were eventually replaced as the conceptual model sharpened.

What remains is an invitation — not to a finished model, but to a field in formation. If SEMT succeeds, it will not be because it answers every question, but because it creates a structured space in which deeper questions can be asked.

This book may be cited freely, distributed openly, and engaged critically. Suggestions, extensions, and critiques are welcome. SEMT is a beginning.

Tribute. This book is quietly indebted to Robert M. Pirsig. His vision of "quality" as the foundational yet unspoken force of structure and value shaped the underlying impulse behind SEMT. Just as he dissolved the false divide between classical and romantic understanding, SEMT dissolves the divide between formal rigor and epistemic evolution. Pirsig showed that structure is not lifeless — it is alive with direction.

This work is a continuation of that insight, translated into mathematical and epistemic terms. The idea, ultimately, is his.

Note on Language and Tools. This book was conceptualized, structured, and authored by Krister Sundholm. Throughout the writing process, I used advanced AI language models as a tool for refining English expression, formatting LaTeX, and stress-testing the coherence of explanations. As English is not my native language, this support allowed me to focus on the structure of thought, rather than syntax. All ideas, definitions, and philosophical arguments originate with the author. The responsibility for the content — and its implications — is entirely my own.

Chapter C

Acknowledgments

This work was supported by moments of insight, extended solitude, and a persistent fascination with structure. I wish to thank those who helped shape the early concepts through conversation, critique, or silent encouragement.

To those thinkers whose unresolved questions made space for SEMT — your work was the invitation. And to future readers who will challenge or extend this framework — your engagement will be its growth.

Finally, thanks to my wife Tina, whose patience, strength, and clarity made this work possible and whose presence grounded even the most abstract layers.

Bibliography

Bibliography

[1] K. Gödel, "Über formal unentscheidbare Sätze der Principia Mathematica und verwandter Systeme I," *Monatshefte für Mathematik und Physik*, vol. 38, pp. 173–198, 1931.

[2] A. M. Turing, "On computable numbers, with an application to the Entscheidungsproblem," *Proceedings of the London Mathematical Society*, vol. 42, no. 2, pp. 230–265, 1936.

[3] S. Kripke, "Outline of a theory of truth," *Journal of Philosophy*, vol. 72, no. 19, pp. 690–716, 1975.

[4] P. J. Cohen, "The independence of the Continuum Hypothesis," *Proceedings of the National Academy of Sciences*, vol. 50, no. 6, pp. 1143–1148, 1963.

[5] G. Chaitin, "Information-theoretic limitations of formal systems," *Journal of the ACM*, vol. 21, no. 3, pp. 403–424, 1974.

[6] M. Fitting, *First-Order Logic and Automated Theorem Proving*, 2nd ed., Springer, 2006.

[7] M. Glanzberg, "Truth," *The Stanford Encyclopedia of Philosophy* (Fall 2021 Edition), E. N. Zalta (ed.).

[8] R. M. Pirsig, *Zen and the Art of Motorcycle Maintenance: An Inquiry into Values*, New York: Harper Perennial Modern Classics, 2006 (original work published 1974).
Note: *Zen* provided the philosophical core for SEMT's structural view of truth and quality. Its influence is foundational and explicit.

[9] R. M. Pirsig, *Lila: An Inquiry into Morals*, New York: Bantam Books, 1991.
Note: *Lila* expands Pirsig's Metaphysics of Quality (MOQ), which parallels SEMT's layered and evolving epistemology. The work's structural ambition and inquiry into valuation directly shaped the theoretical framing of this book.

168 52 Bromma, Sweden
https://aretecopublishing.se

Printed by Libri Plureos GmbH, Friedensallee 273,
22763 Hamburg
ISBN: 978-91-988345-3-6
First edition: May 2025

www.ingramcontent.com/pod-product-compliance
Lightning Source LLC
Chambersburg PA
CBHW070647240926
51846CB00003BA/810
* 9 7 8 9 1 9 8 8 3 4 5 3 6 *